Work 139

U0392560

跳舞的小精灵

Dancing Fairies

Gunter Pauli

[比] 冈特·鲍利 著

[哥伦] 凯瑟琳娜·巴赫 绘

高青 译

上海远东出版社

丛书编委会

主　任：田成川

副主任：闫世东　林　玉

委　员：李原原　祝真旭　曾红鹰　靳增江　史国鹏

　　　　梁雅丽　孟小红　郑循如　陈　卫　任泽林

　　　　薛　梅　朱智翔　柳志清　冯　缨　齐晓江

　　　　朱习文　毕春萍　彭　勇

特别感谢以下热心人士对童书工作的支持：

匡志强　宋小华　解　东　厉　云　李　婧　庞英元

李　阳　梁婧婧　刘　丹　冯家宝　熊彩虹　罗淑怡

旷　婉　王靖雯　廖清州　王怡然　王　征　邵　杰

陈强林　陈　果　罗　佳　闫　艳　谢　露　张修博

陈梦竹　刘　灿　李　丹　郭　雯　戴　虹

目录

Contents

走鹃看着一群沙漠蛛蜂围着马利筋飞来飞去，看来这些发酵的水果让沙漠蛛蜂醉了。

"如果你吃太多的发酵水果会醉的，那我会很容易抓住并吃掉你。"走鹃警告道。

A roadrunner is watching a swarm of large spider wasps buzzing around a milkweed plant. It looks as if feeding on the fermented fruit is making the wasps drunk.

"If you eat too much of this you may get drunk, and it will be very easy for me to catch and eat you," Roadrunner warns.

……发酵的水果让沙漠蛛蜂醉了。

... fermented fruit is making the wasps drunk.

······狼蛛。

... the tarantula.

"你知道吗，你是唯一让我害怕的鸟。"沙漠蛛蜂坦言。

　　"我知道谁最怕你——狼蛛。当它看见你来的时候，它会躲起来，因为它知道你会像鹰一样从天而降，蜇它后还在它的身上放一个蛋，好让你孵出来的宝宝可以以它为食。"

"You know, you are the only bird that I am really scared of," Wasp confesses.

"And I know who is really scared of you – the tarantula. It hides when it sees you coming, as it knows you will drop from the sky like a hawk, sting it and then lay an egg on its body, for your little one to feed on when it hatches."

"嗯，它们为什么叫我狼蛛鹰呢！"

"我不知道为什么，因为你既不是狼蛛也不是鹰。但我知道被你的刺蜇一下真的很痛，是世界上最痛的。"走鹃说。

"Well, they call me the tarantula hawk for nothing!"
"I don't know why, as you are neither a tarantula nor a hawk. What I do know is that your sting really hurts. It must be the most painful in the world," Roadrunner says.

... tarantula hawk ...

……被子弹蚂蚁的刺蜇到更疼。

... the bullet ant's sting hurts even more.

"不对，"沙漠蛛蜂说，"我听说被子弹蚂蚁的刺蜇到更疼。其实给别人造成痛苦并不会带给我任何快乐。我的蓝黑色身体应该足够警告所有人别来惹我了吧……"

"现在你就是惹我的那个，你吃水果吃醉了，在你蜇我之前，我可以迅速抓住并吃了你……"

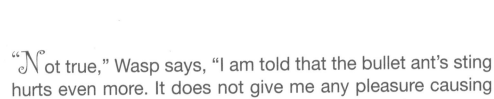

"Not true," Wasp says, "I am told that the bullet ant's sting hurts even more. It does not give me any pleasure causing others pain. My blue and black body should be enough warning to everyone out there not to mess with me …"
"And now you are the one messing about, eating fruit that makes you drunk. I will be able to snatch and eat you long before you can get to sting me …"

"大多数鸟儿都会远离我，但你是例外。你是知道我没有毒液吗？被我蜇了只会感到痛，但不会致命。"沙漠蛛蜂解释道。

　　"我以为是毒液引起的疼痛呢！"

"Most birds stay well away from me, but you are the exception. Do you perhaps know that I have no venom? I only cause pain, I do not kill," Wasp explains.

"I though it was the venom that causes the pain!"

……我没有毒液吗？

... I have no venom?

......给想吃我的家伙们一个严重警告......

... give my predators a serious warning ...

"我只是想给想吃我的家伙们一个它们不会轻易忘记的严重警告。如果它们记得被我蜇过的痛苦，就不会再试图追捕我了。"

　　"哦，我明白了！你只想教训它们一顿。你真的相信你带来的刺痛会让它们记住并远离你吗？"

"All I want is to give my predators a serious warning – one that they won't easily forget. If they remember how painful my sting is, they will not try to hunt me again."

"Oh I see! You only want to teach them a lesson. Do you really believe that your painful sting will help them remember to stay away from you?"

"我相信！你看，我不会造成任何伤害。仅仅三分钟的疼痛过后，一切就都恢复正常了。"

"你的意思是疼痛一会就过去了，然后就……没事了吗？"走鹃问。

"I do! Look here, I don't cause any damage. It takes just three minutes for the pain to pass – and all is back to normal."
"You mean, the pain passes, and then ... nothing?" Roadrunner asks.

... just three minutes for the pain ...

·····沙漠蛛蜂和蜜蜂有很大的区别·····

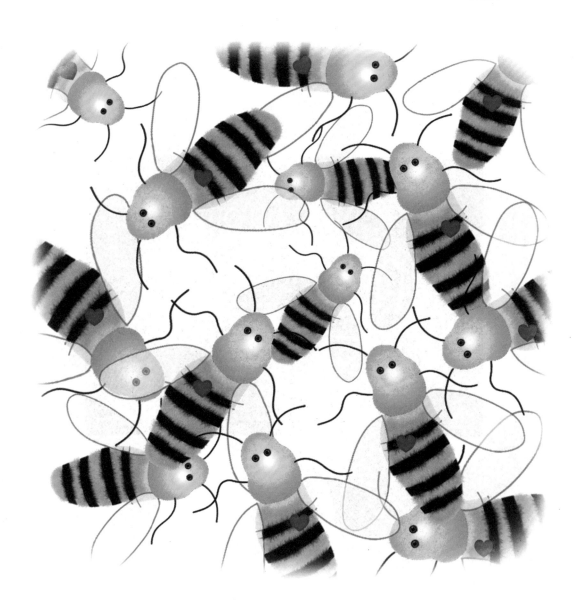

... wasps are very different from bees ...

"没错。疼痛是一种警告信号。它警告身体，伤害已经造成或正在进行中。或者如果你不采取行动，将会造成伤害。而我是不会造成伤害的，但会发出一个严重警告。"

"你欺骗你的天敌的思维方式很有意思。你们沙漠蛛蜂和蜜蜂有很大的区别，蜜蜂通过蜇杀来保护它们的幼虫和蜂蜜。它们不仅造成疼痛，而且被它们蜇刺可能会致命。它们的毒液甚至可以让心脏停止跳动。"

"Exactly. Pain is a warning signal. It warns the body that damage has been done or is being done to it. Or will be done, if you do not take action. In my case, no damage is done, but a strong warning is sent out."

"Interesting, the way you cheat the brain of your predator. You wasps are very different from bees that protect their larvae and honey by stinging to kill. They not only cause pain, but their stings can be fatal. Their venom can even stop the heart from beating."

"好吧，"沙漠蛛蜂说，"你必须区分我们和非洲杀人蜂，它的毒液可以杀死一个人甚至一匹马。和它们相比，我们只不过是跳舞的小精灵而已！"

　　"好的！下一次饿的时候，我会牢记这一点的！"走鹃笑道。

　　……这仅仅是开始！……

"Well," Wasp says, "you have to distinguish between us and killer bees that can kill a human or even a horse. Compared to them, we are no more than just dancing fairies!"

"Good! I will remember that the next time I am hungry!" Roadrunner laughs.

... AND IT HAS ONLY JUST BEGUN! ...

······这仅仅是开始!······

...AND IT HAS ONLY JUST BEGUN! ..

Did You Know?

你知道吗？

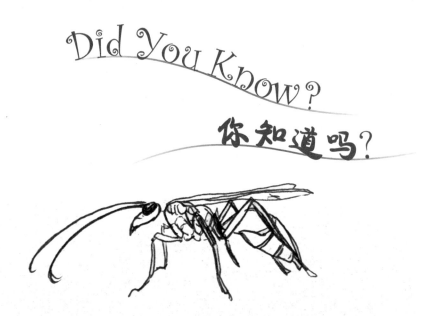

The spider wasp is also called a tarantula hawk because it swoops down on and stings a tarantula, lays one egg on its abdomen, and then covers the tarantula with dirt.

沙漠蛛蜂也叫狼蛛鹰，因为它会俯冲下来蜇刺狼蛛，在狼蛛的腹部产一枚卵，然后用灰尘盖住狼蛛。

If the egg laid on the tarantula is fertilised, a female wasp larva will hatch. If it is an unfertilised egg, the larva will turn into a male wasp.

如果放在狼蛛身上的这枚卵是受过精的，就会孵化出雌性沙漠蛛蜂幼虫。如果它是一枚未受精的卵，就会孵化出雄性沙漠蛛蜂幼虫。

Adult wasps eat nectar and fermented fruit that sometimes intoxicate them to the point that it is difficult for them to fly.

沙漠蛛蜂成虫吃了花蜜和发酵水果，有时会醉得很难再飞起来。

The male wasp waits on a large plant for passing females that are ready to conceive.

雄蜂在一株大的植物上等待准备受孕的雌蜂经过。

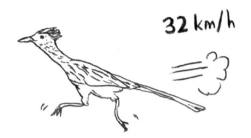

32 km/h

The roadrunner is a bird that prefers running to flying. It can run up to 32 km per hour. This member of the cuckoo bird family occurs in North and Central America.

比起飞翔，走鹃更喜欢奔跑。它的奔跑速度高达每小时32千米。走鹃属于杜鹃鸟家族的成员，生活在北美和中美洲地区。

The roadrunner will reduce its body temperature and the rate of its metabolism to conserve energy during cold nights in the desert.

在沙漠的寒夜里，走鹃会降低其体温和代谢率以保存能量。

The killer bee, also known as the Africanized honeybee, is a hybrid of the Brazilian honeybee and the African bee, which was introduced to Brazil. A swarm of these bees can chase after a person for up to 400 metres.

杀人蜂，又称非洲蜜蜂，巴西引入非洲蜂后，其和巴西蜜蜂杂交产生杀人蜂。一群杀人蜂能追赶一个人长达400米。

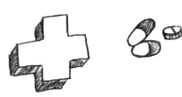

Pain is a symptom and most pain goes away once the cause is removed or the wound has healed. In developed countries, pain is the most common reason for people visiting doctors.

疼痛是一种症状，一旦诱发疼痛的因素去除或伤口愈合，大多数疼痛就会消失。在发达国家，疼痛是人们就医的最常见原因。

Think About It

想一想

Are you scared of pain?

你怕痛吗?

What fermented food is your favourite?

你最喜欢的发酵食品是什么?

Do you remember a moment of great pain? And did you change your behaviour to avoid having to suffer that pain again?

你还记得你经历过的最痛时刻吗? 你是否为了避免再次遭受这种痛苦而改变了你的行为呢?

Are there better ways to teach someone a lesson than by causing him or her intense pain?

有没有一些更好的方法既能教导人们接受教训, 又不会使他或她受到剧烈痛苦?

Have you ever been stung by a bee? Do you remember the pain? Think about some instances when you suffered severe physical pain. Which were the three most painful experiences? Now ask yourself if suffering such pain has made you change your behaviour. Or do you consider that pain to have been the result of an accident, and as a result of it you have not changed your behaviour at all? Now think of at least three instances when you have caused someone else to suffer pain. Did the fact that you were responsible for their pain change your relationship with those people? Reflect on this and discuss your findings with your friends and family.

你被蜜蜂蜇过吗？你还记得那种痛吗？想想你遭受过的严重的身体疼痛。列举三个最痛苦的经历。现在问问自己，当时的这种痛苦是否让你改变了自己后来的行为。或者你认为那时遭受的痛苦只是个意外，你并没有因此改变你的行为？再想想并举出至少三个由于你的原因给别人造成了痛苦的实例。对此你应负担让他们承受痛苦的责任，这一事实是否改变了你和那些人的关系？反思这一点，并与你的朋友和家人讨论你的发现。

学科知识
Academic Knowledge

生物学	走鹃属于杜鹃科鸟类；对趾足：两趾向前，两趾向后；走鹃是杂食动物；同兰花相比，马利筋的花朵更复杂。
化　学	酒精是发酵的糖；葡萄糖和果糖结合形成蔗糖；碳水化合物可分解成葡萄糖；发酵食品包括泡菜、酱油、豆酱、豆豉和酱菜，而发酵饮料包括康普茶、乳清、酸奶、啤酒和葡萄酒。
物　理	当肝脏无法承受进入体内的酒精量时，中枢神经系统就会受到压抑；酒精会阻碍加压素形成，是一种抗利尿激素。
工程学	生物医学工程治疗极端疼痛，这种极端疼痛是长期残疾的最常见的原因之一。
经济学	发酵是一种保存食物的方法，所以发酵食品的保质期较长。
伦理学	让人们记住痛苦的经历从而改变他们的行为，并记住由于他们的行为导致的负面后果。
历　史	莎士比亚说，酒精"激起了欲望，却带走了表演"。
地　理	走鹃生活在北美地区及中美洲。
数　学	渐近曲线。
生活方式	几十年来人们对糖的使用不断增加，现在要提倡减少糖的摄入量；加工食品一般糖含量较高，因为加工时使用果糖、玉米糖浆、蔗糖等。
社会学	某些文化中是容忍醉酒的，另一些文化却是完全禁止醉酒的。
心理学	吃糖是会上瘾的；喝醉时一个人的情绪可以从亢奋到忧郁低落；酒精有镇静作用。
系统论	改变我们行为方法就像生产和消费产品一样是不可持续的。然而，这种变化不应通过痛苦来引导。

情感智慧
Emotional Intelligence

走 鹃

走鹃虽然威胁要吃掉沙漠蛛蜂,但其实是在提醒它。它深知沙漠蛛蜂的攻击性,承认它能致人疼痛。后来走鹃强调了它的威胁,表明它是认真的。开始时走鹃认为沙漠蛛蜂更强大,后来听到它关于毒液和痛苦的解释,走鹃表现出了一些同情。它对沙漠蛛蜂改变别人的方式表示质疑,并怀疑一旦痛苦消失后,行为是否还会发生改变。走鹃认为沙漠蛛蜂是"欺骗",但承认它的方法与杀人蜂的刺是非常不同的,虽然被杀人蜂蜇的痛苦少,但会致命。走鹃承诺记住杀人蜂和沙漠蛛蜂的区别。

沙漠蛛蜂

沙漠蛛蜂是惧怕走鹃的,并承认走鹃是唯一让它害怕的鸟。它一直很谦虚地听走鹃描述它的刺是多么痛,并指出被子弹蚂蚁刺更痛。它又很快补充说,它不喜欢给别人造成疼痛,目的只是警告别人不要惹它。它解释说,它蜇的刺只会引起疼痛,并不含有任何能导致永久性伤害的毒液。它只想让侵犯它的掠食者记住痛苦,还它平静的生活。它想让走鹃区分沙漠蛛蜂和杀人蜂的不同,被杀人蜂蜇虽然不太疼,但其毒液可以杀死人或动物。

艺术
The Arts

让我们找一些表示疼痛的图形符号。虽然很容易画出显示疼痛的面部表情,但还有其他可以显示危险或可能导致疼痛的方法。寻找新的方法并添加到你已经知道的清单里。辨认一些不同文化中选用的与疼痛相关的颜色。将这些颜色添加到你列出的符号表中。这些符号可以表达出你的想法,不必说一个字,也不需使用面部表情。

思维拓展
Systems: Making the Connections

如果我们热衷于回应所有人的基本需求，那么我们目前的生产和消费系统就需要根本性改变。有许多方法可以确保变革发生。第一种方式是改变监管，通过法律的力量。第二种方法是运用税收，惩治破坏环境的行为。还可以借助外在资本，利用公共渠道等。通过有区别的财政体制，提高那些对人们和环境有害产品的征税标准，这样做可以阻止越来越多的人去生产或消费这种产品。第三种方法是确保那些对人和自然都有益的产品，能有快速增长和有保障的市场需求，这使它们能够实现规模经济，降低生产成本，使这些产品在市场上更快地具有竞争力。第四种方法是让人们用非常不同的生产方式生产产品，从而引导消费者转向新的、破坏性低的产品，促使系统转型。这需要创新的技术和商业模式。沙漠蛛蜂通过引起疼痛来根本改变不必要的行为，一旦动物由于不可接受的行为而遭受痛苦，便不可能再去威胁沙漠蛛蜂，因为它会记住自己的行为引发的后果。在人类社会中，我们必须创造多种方式指导社区群众积极转型，使每个人都能受益于所有可用的资源——安全饮用水、营养的食物、舒适的住房、可再生能源、有效的健康护理以及良好的教育。这就是人类应该选择的生活方式：通过不同的途径来追求可持续发展，让所有的人都有过上健康快乐的生活的机会。

动手能力
Capacity to Implement

我们想找到一种方法，让人们改变自己的行为而不必遭受任何痛苦。找出那些仍然是法律允许的，但导致污染的物质或行为。找出那些有如此行为却持有经营许可证的公司。问问自己，停止这种破坏性活动的最好方法是什么？我们的目标是让你有积极和创新性的想法。人们可以想象，通过新的法律法规或额外的税收政策迫使这类公司转型，以使其走向更可持续的模式。或者你可以寻找一种创新方法，彻底改变其业务结构。在这个阶段，你可以自由畅想，并提出一些解决方案，既能为大家提供机会，又不会给任何人带来痛苦。

故事灵感来自

贾斯汀·施密特
Justin Schmidt

　　贾斯汀·施密特是位在杂草、野花和昆虫中长大的美国昆虫学家。他接受了化学专业培训，并写了《野性的刺痛》。为了搜索昆虫叮咬的数据，他走遍了整个美国，收集蚂蚁、蜜蜂和黄蜂的本地物种。多年来他经历了各种被蜇的疼痛，开发了施密特叮咬疼痛指数，并因此在 2015 年获得"搞笑诺贝尔奖"。他在美国农业部工作了 25 年，研究蜜蜂。现在他是亚利桑那大学西南生物研究所的研究员，在那里他研究蜜蜂、黄蜂和蚂蚁生物学以及它们对人类的医学重要性和影响。

图书在版编目（CIP）数据

冈特生态童书.第四辑:修订版:全36册:汉英对照 /
(比)冈特·鲍利著;(哥伦)凯瑟琳娜·巴赫绘;
何家振等译. —上海:上海远东出版社,2023
书名原文:Gunter's Fables
ISBN 978-7-5476-1931-5

Ⅰ.①冈… Ⅱ.①冈… ②凯… ③何… Ⅲ.①生态环
境–环境保护–儿童读物—汉、英 Ⅳ.①X171.1-49

中国国家版本馆CIP数据核字(2023)第120983号
著作权合同登记号图字09-2023-0612号

策　　划　张　蓉
责任编辑　曹　茜
封面设计　魏　来　李　廉

冈特生态童书
跳舞的小精灵
[比]冈特·鲍利　著
[哥伦]凯瑟琳娜·巴赫　绘

高　青　译

记得要和身边的小朋友分享环保知识哦！
八喜冰淇淋祝你成为环保小使者！